S0-DZP-969

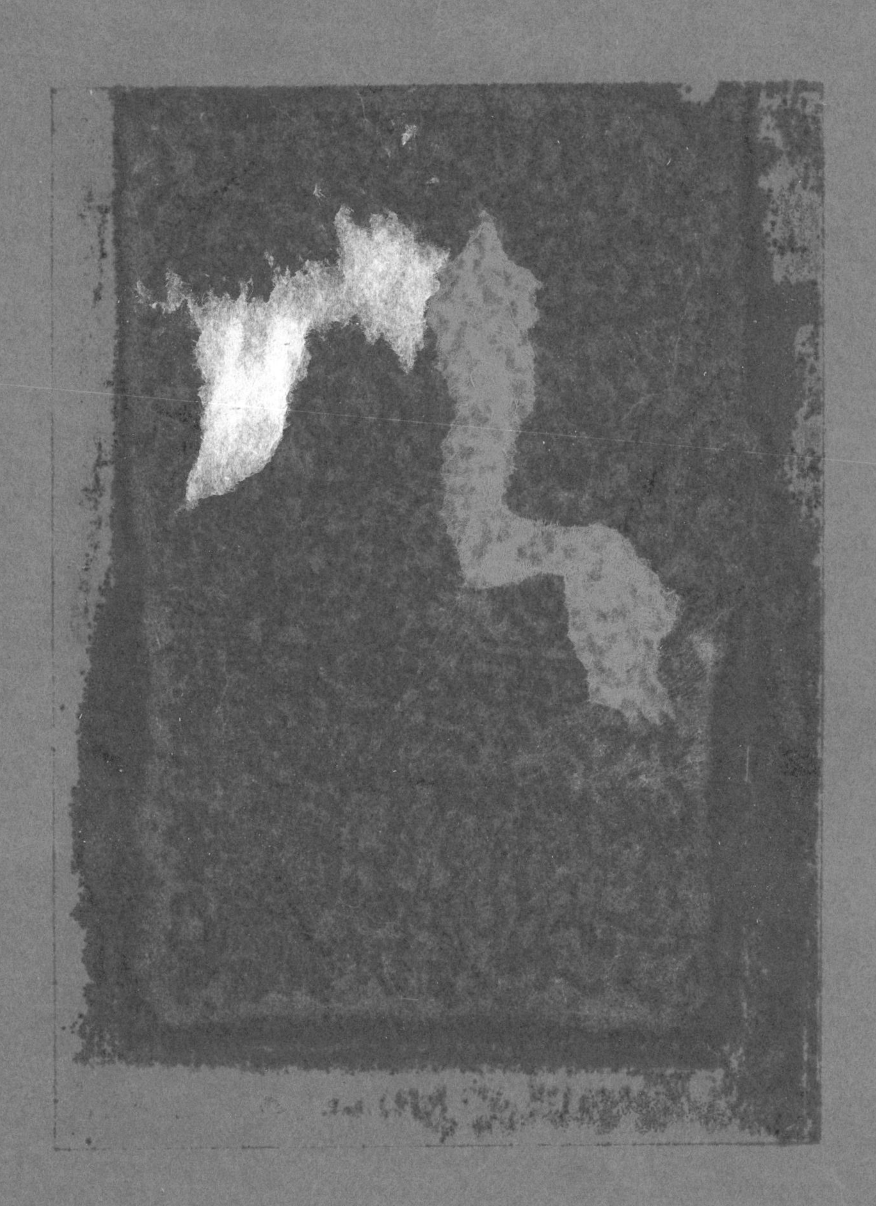

Metamorphosis

METAMORPHOSIS

by Patricia Ryon Quiri

Franklin Watts
New York London Toronto Sydney
A First Book 1991

Cover photograph copyright © Stephen J. Krasemann/DRK Photo

Photographs copyright ©: Frederick D. Atwood: pp. 12, 18 top, 24, 37, 38; Jeff Foott Productions: pp. 16, 18 bottom, 19, 21, 27; DRK Photo: pp. 2, 3 (Michael Fogden), 29 (John Cancalosi), 31 top left, bottom right (Dwight R. Kuhn), 43, 44 (all T. Wiewandt), 49 (Stephen J. Krasemann); Comstock Photography: pp. 31 top right, bottom left (Gwen Fidler), 50 (George D. Lepp).

Library of Congress Cataloging-in-Publication Data

Quiri, Patricia Ryon.
Metamorphosis / by Patricia Ryon Quiri.
p. cm. — (A First book)
Includes bibliographical references and index.
Summary: Describes different types of metamorphosis and details that are undergone by moths, grasshoppers, frogs, marine animals, and others.
ISBN 0-531-20042-6
1. Metamorphosis—Juvenile literature. [1. Metamorphosis]
I. Title. II. Series.
QL981.T47 1991
591.3'34—dc20 91-3104 CIP AC

Printed in the United States of America
6 5 4 3 2 1

To my mother, Jonnie Ryon, for her
never-ending encouragement and support;
and in memory of my late father, Walter Ryon,
who would have been very proud.

CONTENTS

Metamorphosis

INTRODUCTION

Do you enjoy playing outdoors like most boys and girls? If so, you probably have noticed, and maybe even collected, some of nature's little creatures. Have you ever watched a colorful butterfly flutter back and forth among the flowers? Perhaps you've studied a fuzzy caterpillar taking a lazy stroll up the trunk of a tree. Some boys and girls like to follow toads and frogs as they leap from place to place, or examine marine life when they're at the beach. These insects and other animals are among the many that go through a life cycle process called *metamorphosis.*

The word metamorphosis comes from the Greek language. It means "transformation" or "change in form." Some animals that undergo metamorphosis change not only on the outside, but on the inside as well. This metamorphosis can also alter the way in which the animals or in-

sects live. For some, *habitats* and food substances may be different during the various stages of metamorphosis. This is due to the physical transformations and preference changes that take place.

There are different types of metamorphosis. *Entomologists* (scientists who study insects) recognize two kinds of complete metamorphosis: *holometabolous,* which is a typical four-stage metamorphosis, and *hypermetamorphosis,* a four-stage metamorphosis in which the insect looks different between consecutive molts. Complete metamorphosis refers to many insects. It involves the insect going through a total change in body form. This change is accomplished through four stages: egg, larva, pupa, and adult. Insects that completely metamorphose include butterflies, moths, honeybees, mosquitoes, and flies. The young insect looks completely different from the adult insect. The length of time it takes insects to undergo complete metamorphosis varies from species to species. Some insects take as little as several days to metamorphose, while others may take up to three years to completely change form.

Two boys look for grasshoppers to study.

Another type of metamorphosis is simple metamorphosis (sometimes called incomplete). This involves a gradual change in form through three stages: egg, nymph or larva, and adult. *Hemimetabolous* is a simple metamorphosis in which the nymphs are *aquatic* and the adult insects are *terrestrial*. *Paurometabolous* is a simple metamorphosis in which both nymph and adult live in the same habitat. Among the insects that go through a simple metamorphosis are dragonflies and grasshoppers. This metamorphosis is less dramatic than complete metamorphosis. The immature insect looks similar to the adult except that the young one has no wings. The wings and the size of the insect differentiate the adult from the nymph.

A very different type of metamorphosis takes place among frogs, toads, and many marine animals. In these animals, however, metamorphosis is not classified as simple or complete because these terms apply to insects only.

Let's explore the various types of metamorphosis, starting with some insects that go through complete metamorphosis.

Chapter One

COMPLETE METAMORPHOSIS

THE MONARCH BUTTERFLY

An adult female monarch butterfly lays between one hundred and five hundred eggs at a time. She does this on the underside of the leaves of a milkweed plant. This is the only plant on which she will lay her eggs, because this is the only plant her babies will eat. (Other species of butterflies lay their eggs on different types of plant leaves.) The monarch butterfly's eggs are tiny. Within each egg is a yolk. This is food for the developing insect inside the egg. In about four days the eggs hatch. (Other types of butterfly eggs may take up to nine months to hatch.)

After the eggs hatch, the insect is in the second stage of its metamorphosis. It is called the larval stage. The new "babies" don't look anything like butterflies. What hatches from the egg

is a little caterpillar, the larva. The larva usually eats its own eggshell, which is very nutritious. A monarch butterfly larva has black, yellow, and white stripes. It has lots of little legs. At the front end is a hard round head. This caterpillar feeds on the leaves of the milkweed plant. It chews and grinds so much plant material that it grows very rapidly. But its skin is tough and doesn't expand with its growth. When the caterpillar gets too big for its skin, the skin splits and is shed. (The caterpillar has already grown a new skin underneath.) This is called *molting.* Molting occurs a few times. After about a month, the caterpillar stops eating. It is ready to begin the third stage of metamorphosis: the pupal stage.

The larva finds a place to fasten itself—usually the underside of a milkweed leaf or perhaps a twig. It hangs upside down and spins a green case called a *chrysalis* around itself. The chrysalis is dotted with golden specks. The larva no longer crawls or eats. It has changed into a dark-colored pupa. The pupa does not eat. The chrysalis containing the pupa does not move. It

These mating butterflies will produce eggs.

A butterfly egg is the beginning of a butterfly's life—and the first stage of its metamorphosis. After hatching, the larva eats the eggshell. The caterpillar feeds on the milkweed plant, and grows and grows.

The caterpillar then attaches itself to the plant, and prepares to enter the pupal stage. It spins a chrysalis around itself.

hangs for several days. When looking at it, you might not think anything inside could be alive. However, many changes are taking place within the chrysalis. The body of the larva breaks down. The tissues are used to build adult body parts. The pupa develops legs, wings, and the body of the adult insect. When the butterfly is fully developed, the chrysalis splits open. Out comes the mature insect, the beautiful monarch butterfly with its distinctive brown and orange wings and deep black markings. This is the fourth and final stage of metamorphosis, the adult stage.

After emerging, the butterfly instinctively pumps blood into its limp, almost crumpled, wings to make them larger. The wings are wet and must dry out before the insect can fly. No longer does it have lots of legs. No longer does it crawl up and down the milkweed plant, chomping noisily. Its whole body has changed. It has delicate wings used to fly from flower to flower. It has a

At the end of the chrysalis stage, you can see the change of the insect inside. The insect that now emerges from the chrysalis is a beautiful butterfly.

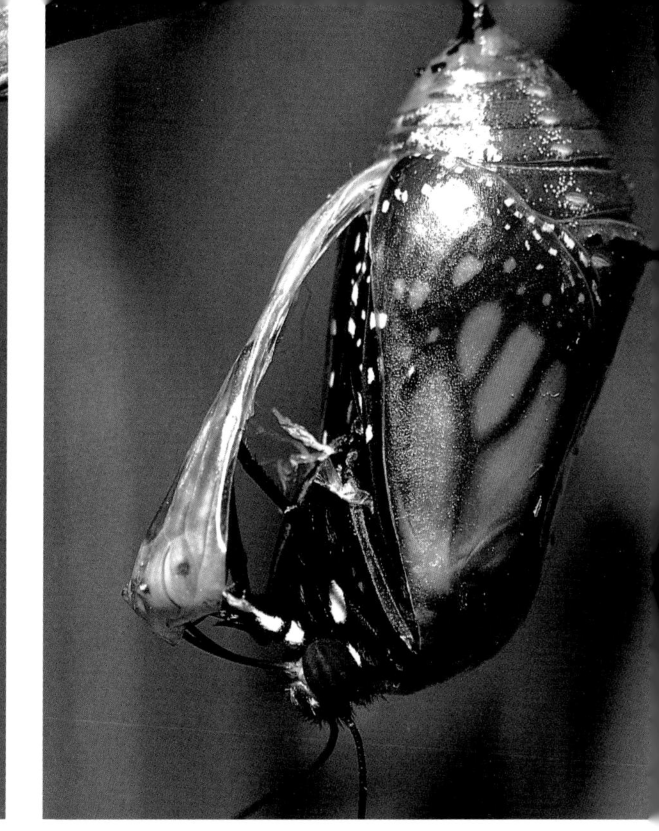

long, thin tube called a *proboscis,* which is a complicated mouthpart. This is used to suck sweet-tasting *nectar* from flowers.

Not only are butterflies pretty and graceful creatures to watch, they also help people in many ways. In the larval stage, the insect eats many plant leaves and in turn fertilizes the soil with its waste material. This makes the soil rich and helps new plants grow. Butterfly larvae also are food for other animals. Because of this, many larvae never have the chance to grow into butterflies. If it does reach adulthood, however, a butterfly helps flowers to reproduce through *pollination.* Pollen from flowers sticks to the butterfly's body as it feeds. When it flies to other flowers searching for more nectar, this pollen brushes off on other flowers. The transfer of pollen is necessary for flowers to reproduce.

The life cycle of a butterfly begins again after a female and a male mate. The cycle repeats itself as the new insects journey through the four stages of complete metamorphosis.

MOTHS

Moths undergo a complete metamorphosis which is similar to that of butterflies. The eggs of moths are also laid on leaves. As with butterflies, the type of plant leaf depends on the particular species of moth. Most moths are highly selective as to where they will lay their eggs, as the plant will become nourishment for their larvae. The eggs hatch into little caterpillars, or larvae. The larva eats the leaves of the plant and grows at a fast rate. It too molts, or sheds its skin. This is done five times. After the second molt, the larva is black and hairy. After the third molt, it is still hairy but has brighter colors. When the final molt is completed, many types of moth larvae leave the plant and hunt for a place on the ground to make their *cocoons.*

The cocoon is similar to a butterfly's chrysalis in that it houses the pupa in the third stage of metamorphosis. A cocoon looks something like a tightly wound cotton ball. The larva spins a covering of silk from its mouth. The newly spun silk, as well as hair from its body, and things found on the ground, such as pieces of leaves and dirt, all make up the cocoon. Inside the cocoon, the skin of the caterpillar splits and a pupa forms. The pupa lies quietly in its little "house" all winter long. But big changes are taking place. The pupa grows wings and legs and the body of a moth. Sometime in the spring the adult moth breaks out of the cocoon. The moth, like the butterfly, must pump its wings full of blood to get them to their mature size. Once they are spread and dry, the adult moth flies away.

Another kind of insect that goes through a complete metamorphosis is the honeybee. Unlike the larva of the butterfly or moth, however, the larva of the honeybee cannot take care of itself. To understand how the honeybee metamorphoses, let's first take a look at how a beehive functions.

There are three types of honeybees in the hive: one queen, hundreds of drones, and thousands of workers. The queen is the largest of the bees. She has only one job, which is to lay eggs. The drones are the male bees. They are smaller than the queen. They also have only one job. Their job is to mate with the queen. The worker bees are the smallest of the three types. Although they are females, they do not lay eggs. These bees come by their name naturally—they are hard workers and are always busy doing various jobs. Have you ever heard the expression "busy as a bee"? Now you know why!

The jobs the workers do depend on their ages.

By the time the moth larva has finished spinning its cocoon, you cannot see through the cottonlike covering.

Some of them take care of the queen by feeding and cleaning her. Others construct honeycomb from wax made in their bodies. Still others gather nectar from flowers and store it in the cells of the honeycomb. The cells are also kept neat and clean by the worker bees. On hot days, some workers cool the hive by waving their wings back and forth, creating "air conditioning." An important job of a young worker bee (about three to ten days old) is caring for the larvae that hatch from the queen's eggs. Here is the story of the honeybee's metamorphosis.

The queen bee mates outside the hive with one of the drones. Back inside the hive she lays between 1,500 and 2,000 eggs a day. She places a little white rice-shaped egg in each of the many cells that make up the honeycomb. The eggs hatch in a few days. This is stage two of the honeybee's metamorphosis, the larval stage. The larva looks like a little white worm and is called a grub. The grub does not leave its cell in search of food. Instead, one of the young worker bees brings food to it. This is quite a job for the worker bee, as the larva eats as often as 1,300

Honeybees gather on their honeycomb.

times a day! The tiny larva eats something called "royal jelly" for the first few days. The worker makes royal jelly in its head! The larva grows bigger and bigger. Then the worker bee feeds it "bee bread." Bee bread is made in other cells of the honeycomb. It consists of pollen and honey.

When the larva reaches the size of its cell, the worker bee makes a wax covering and seals off the cell. This is the third stage—the pupal stage—of the honeybee's metamorphosis. The pupa grows and changes inside the cell for about two weeks. When it is fully developed, the adult form of this insect, the mature honeybee, breaks apart the wax covering and crawls out. It joins the rest of the bees in the honeycomb and gets right to work cleaning the hive.

Bees have important jobs. They furnish us with honey, they pollinate flowers, and they also produce a substance called beeswax. Beeswax is used in making crayons, candles, and cosmetics.

THE MOSQUITO

One probably thinks no further about mosquitoes than what a nuisance they are. These pesky insects come uninvited to picnics and barbecues during the warm weather and can spoil outdoor fun. It is the female mosquito that is the culprit when you get a mosquito bite. She bites people or

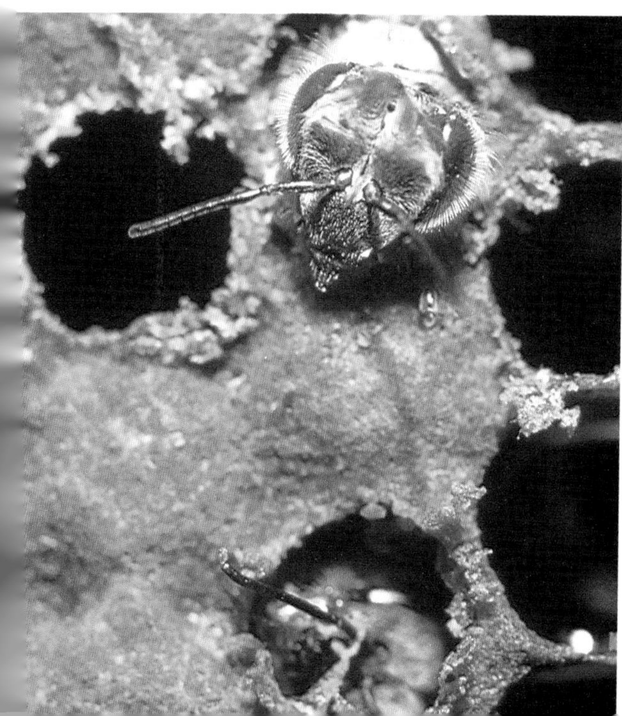

Unlike butterflies and moths, honeybees do not make their own home during the pupal stage. A worker bee seals off the cell for the larva. When the larva has metamorphosed into a mature honeybee, it breaks through the covering.

animals to get her food—your blood or that of another animal. Some female mosquitoes in some parts of the world carry potentially fatal diseases like malaria, yellow fever, and encephalitis (a form of brain fever).

These insects also go through a complete metamorphosis. Have you ever wondered why there are so many mosquitoes near still waters in warm weather? Calm waters are the breeding grounds for these insects. The adult female lays almost one hundred tiny eggs at a time in the water—usually in swamps, creeks, or ponds. Some eggs float together; others float by themselves. They hatch in a few days. One larva comes

Seen here are the four stages of mosquito metamorphosis: eggs clustered together on a pond, larvae hanging beneath the surface of the water, pupae encased during their transformation, and an adult mosquito emerging from its pupal covering.

out of each egg. This larva is very different from the butterfly, moth, or honeybee larva. It lives in the water. Except for a tiny tube for breathing, which stays above the water's surface, the larva of the common mosquito lives beneath the surface of the water. It has two eyes on its head and hangs upside down. Sometimes it wiggles and jerks. The larva eats tiny bits of *microscopic* plants and animals and sheds its covering (called a cuticle) three times during the larval stage, which lasts one to two weeks. It then enters stage three of metamorphosis, and is now a pupa.

The pupa does not eat. Inside the pupal covering, the pupa forms wings. Its intestines change in order to accommodate a strictly liquid diet. The eyes grow bigger. In several days, the case of the pupa rises a bit above the surface of the water and splits open. The adult insect—the mosquito—emerges into the air without getting wet. The life span of an adult mosquito is about five months. However, not many mosquitoes live that long, due to their numerous enemies, including, of course, you!

THE FLY

Another species of often unattractive and pesky insects that undergoes the four stages of metamorphosis is the fly. There are thousands of different types of flies. The familiar types include the common housefly, the horsefly, fruit flies, and gnats. (The mosquito is actually considered a fly as well.)

Flies can sometimes cause harm if they bite. Certain species are carriers of disease. The housefly can spread harmful bacteria which can cause skin irritations as well as intestinal problems. There are some kinds of flies, however, which do help humans. Some get rid of harmful insects. Others help in flower pollination. Also, scientists have used some types of fruit flies in experiments to learn more about genetics, the study of heredity.

All flies go through complete metamorphosis. Female flies lay their small white eggs in many different places: the ground, water, animal waste material, or in rotting matter such as garbage. The eggs hatch into larvae. Depending on the type of adult fly the larvae will become, the larvae are called maggots or grubs. The larvae eat constantly and molt between three and six times. For some flies, the larval stage lasts only a few days; for others, it's as long as a year.

The third stage of metamorphosis—the pupal stage—is a quiet one in the life cycle of most flies. Some pupae are encased in cocoons.

Others, such as the housefly, pupate within the old skin of the larva. When the wings, head, and legs have reached maturity, the adult fly emerges from the pupal covering.

Chapter Two

SIMPLE METAMORPHOSIS

The insects discussed up to this point are just some of those that undergo complete metamorphosis. Many other kinds of insects go through a simple or gradual metamorphosis. There are three stages to simple metamorphosis: egg, nymph or larva, and adult. There is no pupal stage. The changes in simple metamorphosis are not as dramatic as those in complete metamorphosis. However, transformations do take place. Let's take a look at two types of insects that metamorphose simply—dragonflies and grasshoppers.

DRAGONFLIES

An adult dragonfly (sometimes called a "darning needle") has four large wings and a long thin body. Two compound eyes, which consist of many lenses, dominate its rather large head. Bright colors adorn its body and wings. As an adult, this insect can fly extremely fast—up to 60 miles (about 96.6 km) per hour! The dragonfly is *carnivorous,* which means it eats meat. Dragonflies help to keep mosquito, gnat, and fly populations down; those are the insects dragonflies like to eat.

The female dragonfly lays hundreds to thousands of eggs in the waters of quiet ponds, rivers, or swamps. She also might lay them inside the stem of a plant that grows in the water. It takes two to five weeks for most dragonfly eggs to hatch.

After hatching, the insect is in stage two of its metamorphosis, the nymph or larval stage. The dragonfly nymph looks very much like its parent except that it has a shorter and wider body and has yet to develop wings. The brilliant colors of the adult dragonfly are not present in the nymph. The body of the nymph is usually brownish-green. This helps to *camouflage* the nymph in its habitat. Gills located in the rear of its body make it possible for the nymph to breathe in water, where it lives. This second stage of metamorphosis lasts between one and five years. While living as a nymph, the insect nourishes itself by eating smaller insects as well as other insect larvae. Some of the larger nymphs consume tadpoles and very small fish for their meals.

The nymph molts several times before the final stage of metamorphosis. This stage occurs out of the water. The nymph finds something to climb upon, such as a water plant or a branch. It then sheds its skin one last time. Out emerges the

Note that the dragonfly larva looks very much like the adult dragonfly you often see near the water.

An adult dragonfly emerges from its skin. The old skin seen beneath the dragonfly shows that the dragon-fly's metamorphosis is not as extreme as that of some other insects.

adult dragonfly in its splendid colors. Its wings are fully grown. Fortunately for humans, the adult dragonfly feasts upon many pesky insects, but it only does this for about a month, which is the life span for the adult insect.

GRASSHOPPERS

Female grasshoppers lay their eggs in the ground. The eggs usually hatch in the spring, and the baby grasshoppers come crawling out of the ground. This is stage two of simple metamorphosis, the nymph stage. Like dragonfly nymphs, grasshopper nymphs also resemble their parent insect. However, these nymphs are very tiny and are wingless. They have yet to develop reproductive organs. The grasshopper nymph eats lots of grass and green plants. It grows bigger and bigger. When it gets too large for its skin, a hard covering called *chitin,* it molts. Underneath the old chitin is a new and softer skin. This, too, becomes hard for protection. The nymph molts many times, until its wings reach adult size. By the time this happens, its reproductive organs are fully developed. This is stage three of its metamorphosis, the adult stage. As an adult grasshopper, the insect eats the same things it did as a nymph, but now has the ability to fly.

Chapter Three

METAMORPHOSIS IN FROGS AND TOADS

There are three stages in the metamorphosis of frogs and toads: egg, tadpole or larva, and adult. Both frogs and toads are *amphibians*—that means they live in water as well as on land.

Frogs and toads mate in the spring. This takes place in ponds, lakes, or streams. A female frog or toad finds a mate which puts his forelegs around her and climbs on her back. The female's body is heavy with eggs. The two animals swim around until the female lays her eggs. Once the eggs are released from her body, the male fertilizes them. He does this by spraying his sperm on top of them. Depending on the type of frog or toad, the amount of eggs laid can number between one hundred and several thousand. All the eggs are contained in a jellylike substance. The eggs are tiny, round, and dark. Within each fertilized egg, a tadpole starts to grow. It is nourished by food stored in the egg.

After one to three weeks the egg hatches, releasing a tiny tadpole with a round head and a little tail. This is the second stage of the metamorphosis. The newborn tadpole has no eyes and no mouth. For a few days, the tadpole sticks to the egg jelly or a nearby water plant. Its eyes and mouth soon form and its tail gets bigger. Gills grow on the outside of the head. It is through the gills that the tadpole gets oxygen from the water. Wiggling its tail back and forth, the little tadpole is ready to explore.

More often than not, a tadpole makes a tasty meal for hungry insects, fish, or adult frogs. However, the tadpole that survives soon grows gills on the inside of its body while the outside gills disappear. It swims around, finding water plants to feed on. Soon the tadpole grows two hind legs, and shortly thereafter, two forelegs. Gradually, its tail gets shorter and shorter. The tadpole looks more and more like an adult frog or toad. However, something else must develop inside its body so that it can live on the land: lungs. They are necessary for the amphibian to breathe when it is out of the water. The gills soon disappear. During stage two of the metamorphosis, the mouth and intestines change to handle new kinds of food. The eyes get bigger. Within two or three months, the tadpole has metamorphosed into an incredibly tiny frog or toad.

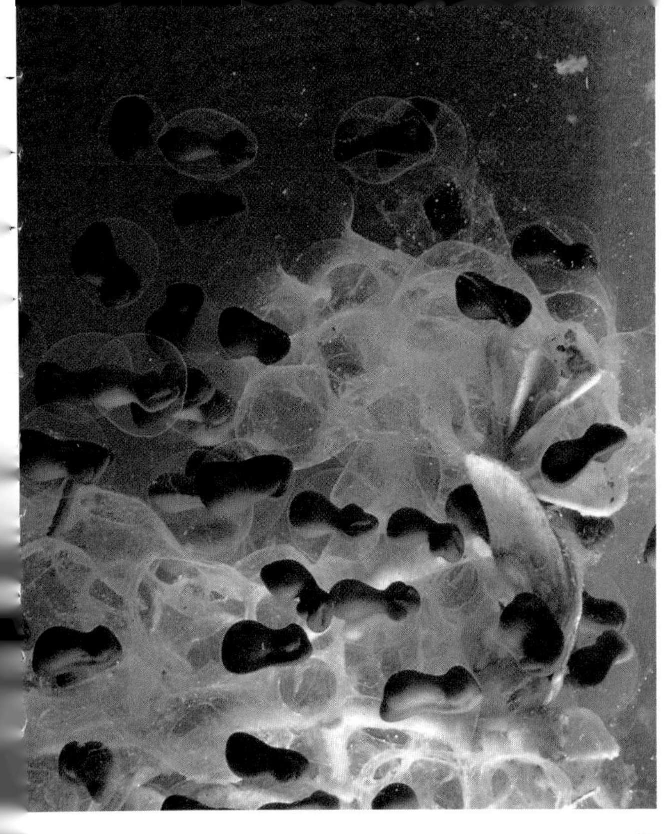

Fifteen hours after they were spawned, these frog eggs (left) have already started to form the shape of the tadpoles they will become (bottom).

As it develops, the tadpole looks more and more like a frog. In the bottom photo, you can see two frogs at different stages of metamorphosis: The one at the left still has a tail!

Life is different in the third stage of metamorphosis for this creature. As an adult it feeds on a variety of insects such as flies, caterpillars, moths, and spiders. Food is trapped by a long, sticky tongue that darts out of its mouth. Adult frogs usually live near water, while adult toads tend to make their homes around bushes, trees, or fields. These amphibians continue to grow. The skin becomes too tight, and eventually it is shed. A new skin has grown underneath.

Adult frogs or toads can live for ten or more years. However, these animals have many enemies in nature. They must be very careful in order to live that long.

Chapter Four

METAMORPHOSIS AND MARINE ANIMALS

There are many types of marine life that undergo metamorphosis. Included among these animals are sea stars, crabs, lobsters, jellyfish, sea anemones, corals, and eels. We will examine a few of them in this book.

SEA STARS

Sea stars, sometimes known as starfish (although they are not fish), usually *spawn* (lay eggs) in the spring. Most sea stars are either male or female. There are some types, however, that produce both eggs and sperm. Depending on the type of sea star, as well as the temperature of the water, the number of eggs laid can range from hundreds to millions.

Days after the eggs are fertilized, they hatch into tiny larvae. The larvae look nothing like adult sea stars. They are bilaterally *symmetrical.* This means they have two sides (like humans), with identical body parts on either side. They are free-swimming. The larval stage lasts anywhere from seven to ninety days. During this period, body parts begin to form that will help the sea stars eat and move. The tiny larvae eat microscopic plants and animals. When the time comes, those larvae that have survived very quickly change into adult sea stars. They sink to the floor of the ocean after developing shells. A stalk grows in the center of the body. The larvae then become radially symmetrical (which means growing out from the center). Within less than twenty-four hours, the larvae become adult sea stars.

All sea stars have a disc in the center of their bodies. Arms (also called *rays*), extend from the disc. The common sea star, among others, has five arms. Other types of sea stars have more than five. The adult moves slowly, using the hundreds of tube feet located in each arm. Clams, oysters, worms, and plants are among the food adult sea stars eat.

CRABS AND LOBSTERS

Female crabs and lobsters carry fertilized eggs under their bodies until it is time for the eggs to

Although all sea stars have discs at their center, you can see the discs of the larger sea stars more clearly in this photo.

Each lobster egg, at left, will hatch into a larva as seen below.

hatch. Many adult females then scatter their eggs into the sea. The newly hatched larvae swim on the surface of the water, eating tiny plants and animals. The larvae are transparent—you can actually see through their small bodies.

Molting is the means by which these animals grow. After shedding their tiny shells, new shells grow to accommodate the changing animal. As time goes by, the animal's shell becomes too heavy for its inhabitant to live on the water's surface. The animal then sinks to the bottom of the sea and begins to function as an adult crab or lobster.

Chapter Five

WHY DOES METAMORPHOSIS OCCUR?

Scientists think hormonal secretions play a big part in the metamorphosis of insects and other animals. *Hormones* are chemicals which stimulate changes in the body. They are released by glands and carried throughout the body by the blood.

For example, the *thyroid* and *pituitary* glands regulate the tadpole's changes to adult frog or toad. The pituitary gland, which is in charge of growth, sends a signal to the thyroid gland. The thyroid then produces a hormone called *thyroxin.* When thyroxin passes through the blood, the tadpole undergoes its metamorphosis. If a tadpole's thyroid gland is taken out, the tadpole will never change into a frog. Some scientists believe the changing seasons stimulate the hormone secretions.

The eggs of most insects, amphibians, and marine animals are tiny. Considering this, it is no wonder the adult insect or animal does not directly hatch from the egg. It takes much growth and time to reach the final stage of metamorphosis. This is good because the competition between the larvae and the adult forms is reduced. Most larvae feed on food substances different from those the adults feed on. In many instances, the habitats of the larvae and the adults are also not the same. Because of these differences in food preference and habitats, the environment can easily accommodate these insects and animals in their various life-cycle stages.

Metamorphosis in insects, amphibians, and marine animals is a complicated and amazing process. The next time you watch a delicate butterfly fluttering from flower to flower, or study a munching caterpillar taking a lazy stroll, or follow a frog hopping from place to place, or sight a crab along the shore, you might be filled with awe. You will know that these small creatures undergo very complex life cycles! Metamorphosis is truly one of nature's many miracles.

Chapter Six

ACTIVITIES, OBSERVATIONS, AND EXPERIMENTS

You may be interested in observing some of nature's creatures in the various stages of metamorphosis. Listed below are some activities you might enjoy.

NATURE WALKS

Go to the park or roam around outside your home. Check leaves for insect eggs and larvae. Leaves with holes indicate something has been feeding on them. Place eggs or larvae, plus the leaves, in a jar. Cap the jar and poke holes in the lid. Keep the jar well supplied with leaves. Record your observations daily. Free the insect once it has metamorphosed into an adult.

If you don't want to collect eggs or larvae, it is fun just watching a caterpillar move and eat. Listen closely. You may hear it munching. Record the spot in which you first observed the caterpillar. Go back the next day and see if it is still there.

TADPOLE-WATCHING

If you live near a pond, perhaps you'd like to look for frog eggs. Do this in early spring. Fill a pail half-full with pond water. Scoop a small number of eggs into the pail. Put in some water plants. After the eggs hatch, keep only a few tadpoles in your pail. Return the others to the pond. When the tadpoles get bigger, put live insects and worms in the pail as well as more water plants. Add pond water as needed. Place a large rock in the pail. The larger tadpoles may want to rest on it when they breathe above the water. Record the tadpoles' growth. After they metamorphose into young frogs, return them to their natural environment.

FOR FURTHER READING

Abels, Hariette S. *Killer Bees.* New York: Crestwood, 1987.

Bailey, Jill. *Discovering Crabs and Lobsters.* New York: Bookwright Press, 1987.

Coldrey, Jennifer. *Frog in the Pond.* Milwaukee: Gareth Stevens, 1987.

Fischer-Nagel, Heiderose and Andreas. *Life of the Butterfly.* Minneapolis: Lerner Publications, 1988.

Hornblow, Leonora and Arthur. *Insects Do the Strangest Things.* New York: Random House, 1968.

Oda, Hidetomo. *Observing Bees & Wasps.* Milwaukee: Raintree, 1986.

Patent, Dorothy H. *Mosquitoes.* New York: Holiday, 1986.

Webster, David. *Frog & Toad Watching.* New York: Messner, 1986.

GLOSSARY

Amphibian—an animal that is able to live in water as well as on land.

Aquatic—living in water.

Camouflage—the way in which insects or animals blend in with their background.

Carnivorous—flesh-eating.

Chitin—hard covering of skin in insects.

Chrysalis—the case in which the pupa of a butterfly rests.

Cocoon—the case in which the pupa of a moth rests.

Entomologist—a specialist in the study of insects.

Habitat—natural home of an insect or other animal.

Hemimetabolous—a type of simple metamorphosis in which the nymph lives in water and the adult lives on land; adult insects have wings.

Holometabolous—typical, four-stage metamorphosis.

Hormones—chemical-like substances, which create change when released in the bodies of insects and other animals.

Hypermetamorphosis—a four-stage metamorphosis in which the insect looks different between consecutive molts.

Microscopic—very tiny; can only be seen with a microscope.

Molt—to shed the old skin.

Nectar—sweet liquid produced by flowers.

Paurametabolous—a type of simple metamorphosis in which the nymph and adult live in the same habitat; adult insects have wings.

Pituitary gland—body gland responsible for growth.

Pollination—the way in which flowers reproduce by the transfer of pollen from flower to flower.

Proboscis—a complicated mouthpart in some insects, used for sucking liquid food from flowers.

Rays—the arms of a sea star.

Spawn—to produce or lay eggs.

Symmetric—having balanced proportions.

Terrestrial—living on land.

Thyroid gland—body gland which regulates growth and development.

Thyroxin—a hormone produced by the thyroid gland.

INDEX

Page numbers in *italics* refer to illustrations

ABOUT THE AUTHOR

Patricia Ryon Quiri lives in Palm Harbor, Florida, with her husband and three sons. She is a former elementary school teacher and has an elementary education degree from Alfred University in New York. Other Franklin Watts books by Ms. Quiri include *Dating* and *Alexander Graham Bell.*